Cambridge Primary
Mathematics

Second Edition

Workbook 5

Steph King
Josh Lury
Series editors:
Mike Askew
Paul Broadbent

T0173180

Boost

HODDER
EDUCATION
AN HACHETTE UK COMPANY

Cambridge International copyright material in this publication is reproduced under licence and remains the intellectual property of Cambridge Assessment International Education.

Third-party websites and resources referred to in this publication have not been endorsed by Cambridge Assessment International Education.

Registered Cambridge International Schools benefit from high-quality programmes, assessments and a wide range of support so that teachers can effectively deliver Cambridge Primary. Visit www.cambridgeinternational.org/primary to find out more.

Acknowledgements

The Publishers would like to thank the following for permission to reproduce copyright material.

Photo credits

p. 8 *tl, cr,* **p. 14** *tl, cr,* **p. 21** *tl, cr,* **p. 24** *tl, cr,* **p. 30** *tl, cr,* **p. 36** *tl, cr,* **p. 41** *tl, cr,* **p. 45** *tl, cr,* **p. 52** *tl, cr,* **p. 57** *tl, cr,* **p. 63** *tl, cr,* **p. 68** *tl, cr,* **p. 73** *tl, cr,* **p. 76** *tl, cr,* **p. 83** *tl, cr,* **p. 87** *tl, cr,* **p. 92** *tl, cr,* **p. 96** *tl, cr* © Stocker Team/Adobe Stock Photo.

t = top, *b* = bottom, *l* = left, *r* = right, *c* = centre

Every effort has been made to trace all copyright holders, but if any have been inadvertently overlooked, the Publishers will be pleased to make the necessary arrangements at the first opportunity.

Hachette UK's policy is to use papers that are natural, renewable and recyclable products and made from wood grown in well-managed forests and other controlled sources. The logging and manufacturing processes are expected to conform to the environmental regulations of the country of origin.

Orders: please contact Hachette UK Distribution, Hely Hutchinson Centre, Milton Road, Didcot, Oxfordshire, OX11 7HH. Telephone: +44 (0)1235 827827. Email: education@hachette.co.uk Lines are open from 9 a.m. to 5 p.m., Monday to Friday. You can also order through our website: www.hoddereducation.co.uk

ISBN: 978 1 3983 0122 1

© Steph King and Josh Lury 2021

First published in 2017

This edition published in 2021 by

Hodder Education,

An Hachette UK Company

Carmelite House

50 Victoria Embankment

London EC4Y 0DZ

www.hoddereducation.com

Impression number   10 9 8 7 6 5 4
Year   2025 2024 2023

Cover illustration by Lisa Hunt, The Bright Agency

Illustrations by James Hearne, Stéphan Theron

Typeset in FS Albert 12/14 by IO Publishing CC

Printed in the UK

A catalogue record for this title is available from the British Library.

Contents

Unit 1 — Number

Can you remember?

Circle the numbers that have a value larger than 45 300.

44 900	45 301	4 595
46 100	45 035	45 299

Introducing decimal numbers

1 Draw lines to match the numbers.

1.8	zero point four
18.1	forty-three point nine
0.4	four point three
4.3	one point eight
43.9	eighteen point one

2 Write the value of the underlined digit in each number. Work down each column.

a 2̲5.2   **20**

   25.2̲   ☐

   252.̲5   ☐

b 70.8̲   ☐

   78̲.3   ☐

   78̲3.1   ☐

c 0.7̲   ☐

   67̲.7   ☐

   77.6̲   ☐

Composing, decomposing and regrouping

 1 Look at the example. Decompose the remaining numbers in a similar way.

323 = 300 + 20 + 3   **a** 32.3 =   **b** 33.2 =

c 17.9 =   **d** 97.1 =   **e** 190.7 =

 2 Regroup the number at the top of the regrouping wall into the number of parts in each row. Check that each row is still equal in value to 547 628.

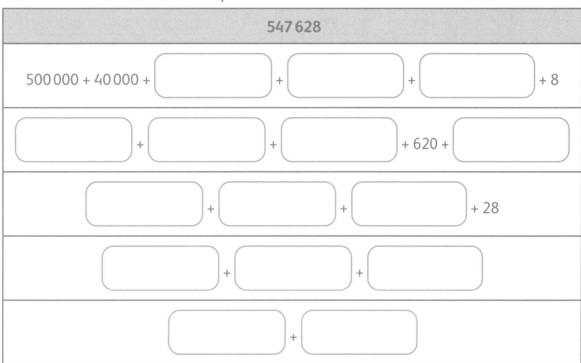

547 628

500 000 + 40 000 + ⬚ + ⬚ + ⬚ + 8

⬚ + ⬚ + ⬚ + 620 + ⬚

⬚ + ⬚ + ⬚ + 28

⬚ + ⬚ + ⬚

⬚ + ⬚

Multiplying and dividing whole numbers by 10, 100 and 1 000

 1 Complete the calculations by filling in the missing numbers. Work down the columns.

a

175 × 100 = ⬚

17 500 ÷ ⬚ = 175

175 × 1 000 = ⬚

175 × 100 × ⬚ = 175 000

b

2 340 ÷ 100 = ⬚

2 345 ÷ ⬚ = 234.5

2 345 × 100 = ⬚

234 500 ÷ ⬚ = 234.5

2 1 000 times as many spectators can watch a football match at Venue A than can watch a swimming event at Venue B.

The number of possible spectators at Venue B is 10 times as small as the number of people who can watch a basketball game at Venue C.

Venue C holds between 5 500 and 5 750 spectators.

Fill in the table to show some possible numbers of spectators at each venue.

	Spectators	Spectators	Spectators	Spectators	Spectators	Spectators
Venue A						
Venue B						
Venue C						

Counting on and back

1 You need a paperclip and a pencil to use with your spinner.
Spin the spinner to make four counting patterns, each with a different step size.
The third number in your count is always 31.
Write four different counts and the step size you used.

_____ , _____ , 31, _____ , _____

Step size:

_____ , _____ , 31, _____ , _____

Step size:

_____ , _____ , 31, _____ , _____

Step size:

_____ , _____ , 31, _____ , _____

Step size:

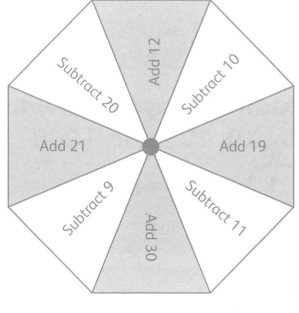

2 Use the numbers to make up different counts, each with five numbers.
You may use the numbers more than once.

-14   7   4   13   -11

-5   1   -23   25

a Count back in sixes: _____

b Count on in nines: _____

c Count back in twelves: _____

Linear sequences

1 Find the rule and the values of the missing terms.

a _____, 48, _____, _____, 24, _____, 8   Rule:

b 2.3, _____, 7.3, _____, _____, _____   Rule:

c 100, _____, _____, 430, _____, _____   Rule:

2 Look at this pattern of sequences. The steps in each pattern are the same size.
a Find the rule and the values of the missing terms.

10, 130, _____, _____, _____, _____   Rule:

10, _____, 130, _____, _____, _____   Rule:

10, _____, _____, 130, _____, _____   Rule:

b What is the next sequence and rule in this pattern?

10, _____, _____, _____, _____, _____   Rule:

Unit 1   Number

Self-check

 I can do this.

 I can do this, but I need to keep trying.

 I can't do this yet.

See how much you know!

What can I do?			
1 I can say and explain the value of different digits in a numeral, for example: the 4 in 86.4 has the value 4 tenths.			
2 I can explain the values of digits in numerals where digits are repeated, for example: the first digit 1 in the number 102.1 represents 1 hundred, whereas the second digit 1 represents 1 tenth.			
3 I can read and write numbers (symbols and words) with up to one decimal place.			
4 I can decompose numbers into the related place value numbers, including decimals, for example: 20.5 = 20 + 0 + 0.5.			
5 I can regroup numbers in a variety of ways, for example, I can express 20.5 as: 20 + 0.5, 20 ones and 5 tenths, 205 tenths.			
6 I can multiply and divide whole numbers by 10, 100 and 1 000 and explain the answers using place value.			
7 I can read and write integers that are less than 0.			
8 I can count on and back in equal steps, including using negative numbers, for example, count back from 15 in sevens: 15, 8, 1, −6, ….			
9 I can position positive and negative numbers around zero, recognising that negative numbers are to the left of zero on a number line.			
10 I can read and say −34 as 'negative thirty-four' (not 'minus thirty-four').			
11 I can identify missing numbers in linear sequences, for example: 11, __, __, 23 = 11, **15**, **19**, 23.			

I need more help with:

_____

_____

_____

Can you remember?

Fill in the table to describe the properties of each shape.

2D shape	Square	Hexagon	Octagon	Pentagon	Triangle
Number of sides					
Number of vertices (corners)					

Symmetrical patterns

1 Shade two more squares to make each pattern symmetrical.
Draw the lines of symmetry.

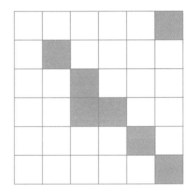 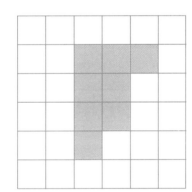 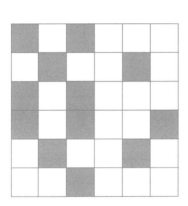

2 Shade squares to make symmetrical patterns.
Make sure that at least one design has diagonal lines of symmetry.
Draw the lines of symmetry.

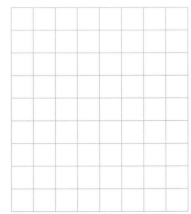 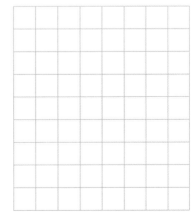 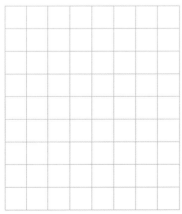

3   Below are four games for two players at a time. The aim is to make a symmetrical pattern on each grid. Taking turns, Player 1 shades one square on the grid. Player 2 then shades a square that is symmetrical.

Shade half squares to challenge yourselves!

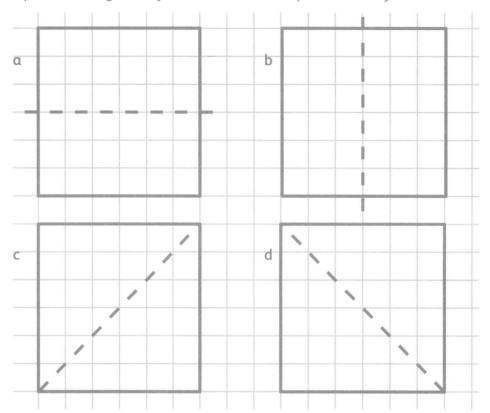

a

b

c

d

Identifying and reasoning about angles

1   Label each angle on each shape:
     A for acute   **O** for obtuse   **R** for reflex.

a

b

c

d

2   **a** Predict the number of different angles for each shape.
Record your predictions in this table.

Shape	Number of acute angles	Number of right angles	Number of obtuse angles	Number of reflex angles
Kite				
Triangle				
Pentagon				
Hexagon				

b Draw an example of each shape on the grid below. Were your predictions correct?

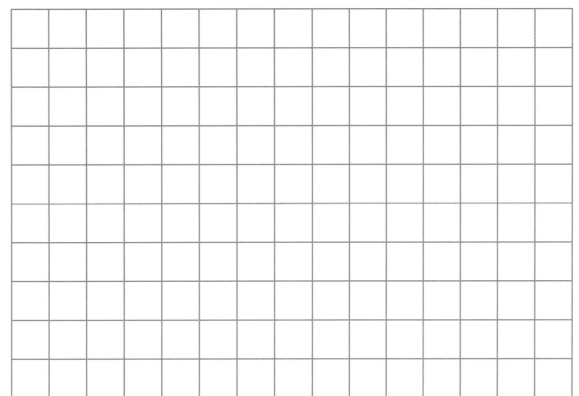

3   Calculate the missing angles. Then name them **acute** or **obtuse**.

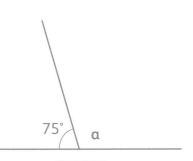

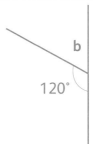

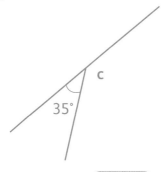

Angle **a** = [ ]°

Angle **b** = [ ]°

Angle **c** = [ ]°

Name: _____

Name: _____

Name: _____

4 Predict which missing angles are a multiple of 5.
Put a **P** next to these. Then calculate to check.

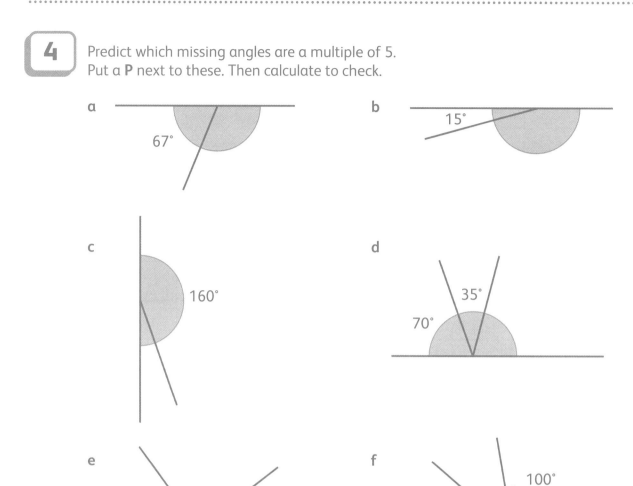

a 67°

b 15°

c 160°

d 35° 70°

e 37°

f 100° 40°

Triangles

1 Each line below is one side of an isosceles triangle. Use a ruler to measure it.
Finish drawing each isosceles triangle. Mark the equal lengths.
Then sketch the lines of symmetry and label any equal angles.

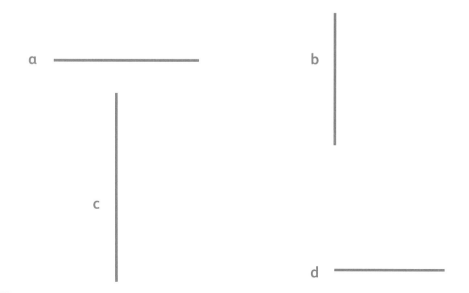

a

b

c

d

2 Under each heading, draw two different triangles of that type. Mark any equal lengths.

a   Equilateral   b   Isosceles   c   Scalene

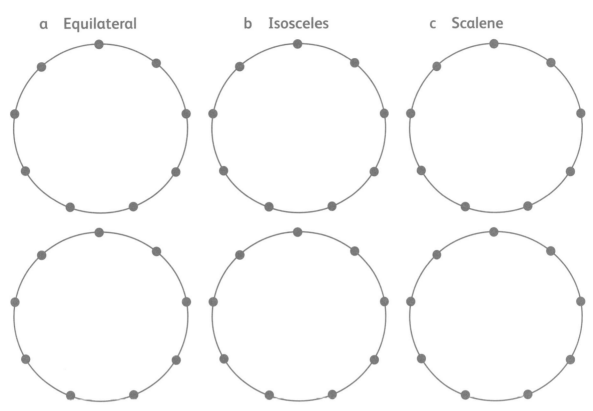

3 Draw a line to split each quadrilateral into two triangles. Name each type of triangle.

a

_____

b

_____

c

_____

d

_____

Unit 2   Angles and shapes

Self-check

 I can do this.

 I can do this, but I need to keep trying.

 I can't do this yet.

See how much you know!

What can I do?			
1 I can identify lines of symmetry in designs and patterns.			
2 I can complete symmetrical designs and patterns.			
3 I can identify, name and label acute, right, obtuse and reflex angles.			
4 I can find the missing angle on a straight line.			
5 I can identify, name and describe the properties of isosceles, equilateral and scalene triangles.			
6 I can find equal lengths and angles in triangles.			

I need more help with:

Can you remember?

Write the related multiplication fact and the two division facts each time.

6 × 5 =	4 × 8 =	7 × 6 =	9 × 4 =

Calculating with positive and negative numbers

1 Use the number line to help you work out these additions.

−15 −14 −13 −12 −11 −10 −9 −8 −7 −6 −5 −4 −3 −2 −1 0 1 2 3 4 5 6 7 8 9 10 11 12 13 14

a −14 + 6 = ☐

   −13 + 7 = ☐

   −12 + 8 = ☐

b −4 + 6 = ☐

   −3 + 7 = ☐

   −2 + 8 = ☐

c ☐ + 12 = 3

   ☐ + 14 = 7

   −5 + ☐ = 11

Describe any patterns you notice in parts **a** and **b**.

_____

_____

2 Complete these subtractions. Use the number line in question 1 to help you.

a 6 − 14 = ☐

   7 − 13 = ☐

   8 − 12 = ☐

b 4 − 16 = ☐

   3 − 17 = ☐

   2 − 18 = ☐

c ☐ − 12 = −3

   ☐ − 14 = −7

   5 − ☐ = −11

Describe any patterns you notice in parts **a** and **b**.

_____

_____

3 The thermometer in the table shows the temperature at the start of the day.
Draw a line graph below to show the information in the table about temperatures.

Time	06:00	10:00	14:00	18:00	22:00
Temperature		Temperature rises 8 degrees	Temperature rises 11 degrees	Temperature drops 5 degrees	Temperature drops 9 degrees

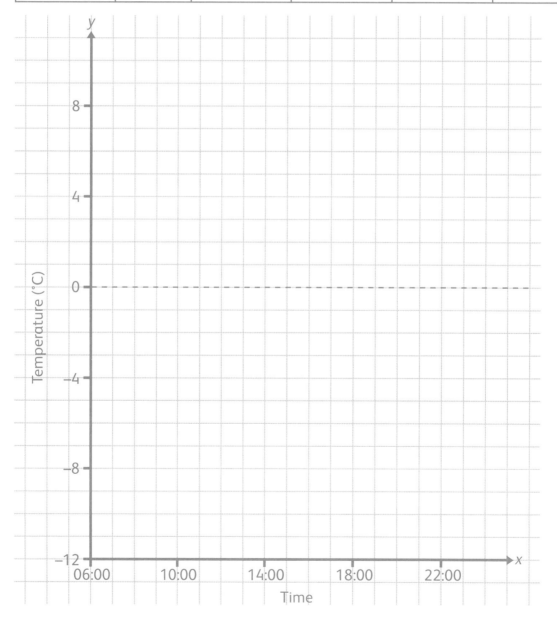

Addition and subtraction

1 Choose any of the three strategies to solve each calculation.
Use each strategy at least once.

Regrouping   Reordering   Decomposing

Calculation	Show your working
1 963 + 145 + 37	1 963 + 145 + 37 = = 2000 + 145 = 2145
a 4 672 – 199	
b 2 439 – 723 – 1 439	
c 450 + 199 + 101 + 550	
d 1 638 + 249	
e 3 568 – 234	

2 Tourists can choose to go on the Red route or the Green route for a sightseeing tour around a city. Each route is split into shorter parts.

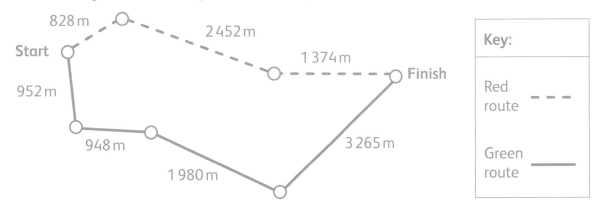

How much longer is the Green route than the Red route?

3 Use estimates to check each calculation. Then calculate to see if you are right. What methods will you use?

	Estimate	Answer ✓ or ✗	Calculation
a	3 694 + 2 147 = 6 841			
b	5 248 – 2 987 = 2 161			
c	6 972 – 843 – 49 = 6 180			
d	609 + 488 + 1 125 = 2 222			

Missing number problems

1

I bought two notebooks and a pen for $7.

I bought the same notebooks as Banko. Three were $4.50.

What is the cost of each item?

a One notebook $ _____   b One pen $ _____

2 Find the value of the symbols in these missing number problems.

a  – 70 = ★

  ★ + ★ + ★ = 210

   = _____

  ★ = _____

b 240 + ⬡ + ⬡ = ▱

  1 000 – ▱ = 300

  ⬡ = _____

  ▱ = _____

c △ + ⬠ = 60

  △ + △ + △ + △ = 100

  △ = _____

  ⬠ = _____

Simplifying multiplications

1 a

$300 × 5 =$ _____

$3 × 5 =$ _____

$50 × 30 =$ _____

$500 × 30 =$ _____

b

$300 × 4 =$ _____

$3 × 4 =$ _____

$400 × 3 =$ _____

$30 × 40 =$ _____

2 Jin multiplies each number by eight. He uses factors to help him.

a **Critique** Jin's work. Which calculations are correct?

	Tick (✓)
$15 × 8 = 15 × 2 × 5 = 30 × 5 = 150$	☐
$35 × 8 = 35 × 2 × 4 = 70 × 4 = 250$	☐
$24 × 8 = 24 × 2 × 2 × 2 = 48 × 2 × 2 = 96 × 2 = 192$	☐
$18 × 8 = 2 × 9 × 8 = 64 × 2 = 128$	☐
$40 × 8 = 10 × 4 × 8 = 10 × 36 = 360$	☐
$25 × 8 = 25 × 4 × 2 = 100 × 2 = 200$	☐

b **Improve** any calculations that are incorrect.

_____

_____

_____

_____

Multiplying numbers up to 1000

1 Work out each answer using one of the two methods.
Write the calculation and answer each time under the method you choose.

Mental method		Written method
25 × 16 = 400 ◄——————— ⎯ 25 × 16		
	34 × 20	
	199 × 7	
	400 × 9	
	460 × 2	
	278 × 6	
	34 × 26	

2 Each day for 48 days, Sanchia saves 75 cents.
Each week for 9 weeks, Pia saves 345 cents.

a Who saved the most money? _____

_____

b How much more did the person in part **a** save? ⬚ cents

3 Use two odd digits and two even digits to make this calculation correct.

⬚⬚ × ⬚⬚ = three-digit odd number

Use the same four digits to make the next calculation correct.

⬚⬚⬚ × ⬚ = four-digit even number

Find four possible digits that you can use. Show your calculations.

_____

_____

_____

_____

Unit 3   Calculation

Self-check

 I can do this.

 I can do this, but I need to keep trying.

 I can't do this yet.

See how much you know!

What can I do?			
1 I can add a positive integer to a negative integer, for example: $-7 + 18$.			
2 I can subtract a positive integer from an integer that has a negative answer, for example: $9 - 12$.			
3 I can use negative numbers in context, for example: a temperature of $3\,°C$ that drops to $-5\,°C$ has fallen by 8 degrees.			
4 I can decide whether to work mentally, with jottings or using a formal method, for example: $243 + 171 + 357 = 243 + 357 + 171 = 600 + 171$			
5 I can use rounding to estimate answers and say whether I have overestimated or underestimated.			
6 I can say when an answer is incorrect by estimating, for example: $498 - 198$ cannot be 296 because we are subtracting less than 200.			
7 I can use decomposing and/or regrouping to make calculations easier, for example: $3\,285 - 1\,495 = 3\,285 - 1\,500 + 5$			
8 I can recognise and use symbols or shapes to represent unknown quantities.			
9 I can use related facts and inverse operations to help with some missing number problems.			
10 I can use the laws of arithmetic to help me simplify calculations.			
11 I can estimate and multiply whole numbers up to $1\,000$ by 1-digit or 2-digit whole numbers.			
12 I can apply skills of calculating and known facts to problems.			

I need more help with:

_____

_____

_____

Unit 4 — Time

Can you remember?

Draw the hands on the clocks to show these times:

a 10:45

b 20:45

c 17:55

d 07:55

Measuring time

1 Sort these times from shortest to longest. Write only the letters.

a 45 seconds   **b** 0.5 hours   c 2.5 minutes   **d** 25 minutes

e 50 minutes   **f** 1 minute   g 100 minutes   **h** 1.5 hours

Shortest ☐ ☐ ☐ ☐ ☐ ☐ ☐ ☐ Longest

2 Draw lines to match the equal times.

0.5 hour	60 hours
0.5 minute	90 seconds
0.5 day	150 minutes
1.5 hours	30 minutes
1.5 days	12 hours
1.5 minutes	90 minutes
2.5 hours	150 seconds
2.5 days	30 seconds
2.5 minutes	36 hours

3 Draw or describe an event for each estimated length of time.

1.5 weeks	1.5 seconds
1.5 years	1.5 months

Calculating time intervals

1 The poster shows the opening times of the swimming pool.

Monday	07:30–20:30
Tuesday	07:00–20:30
Wednesday	07:00–20:30
Thursday	07:30–21:30
Friday	07:00–21:30
Saturday	08:00–19:30
Sunday	08:00–17:45

a For how long is the swimming pool open on Monday? _____

b For how much longer is the pool open on Friday than on Tuesday? _____

c For how many hours is the pool open in total in one week? _____

2 Here are the times for some television programmes.

11:20	A	Wildlife Show
12:00	B	Questions and Answers Quiz
13:15	C	Music Life
14:05	D	Beautiful Gardens
15:30	E	The Art Show
16:50	F	News

a Write television programmes A, B, C, D and E in order of duration from shortest to longest.

Shortest ☐ ☐ ☐ ☐ ☐ Longest

b David has 120 minutes of spare time to watch two programmes. Which two programmes are exactly 120 minutes in total?

_____

_____

3 The clocks show two times on the same day. Write the time that is exactly halfway between the two times.

☐ ____ : ____ ☐

23:20

p.m.

Unit 4　Time

Self-check

 I can do this.

 I can do this, but I need to keep trying.

 I can't do this yet.

See how much you know!

What can I do?	😉	😐	🙁
1　I can understand time that is written in decimals.			
2　I can estimate how long an event takes.			
3　I can solve problems using 12-hour and 24-hour times.			
4　I can calculate the difference between two times.			

I need more help with:

_____

_____

_____

Can you remember?

Complete the table about favourite books.

	Story book	Science and nature book	Picture book	Total
School A	120	80	100	
School B	165	140		420
School C	199		101	410

Bar charts and dot plots

1 The table on the right shows data about the number of people with birthdays on each day of the week in 2020. Use the information to complete the bar chart below. One day has been done for you.

Remember to fill in the *x*-axis and *y*-axis, as well.

Number of people with birthdays on each day of the week in 2020

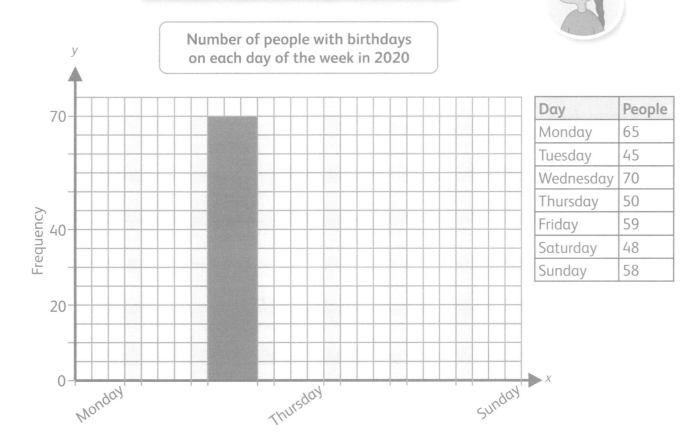

Day	People
Monday	65
Tuesday	45
Wednesday	70
Thursday	50
Friday	59
Saturday	48
Sunday	58

Day of the week

2 Two shops record the number of items they sell during six months.

Shop A	
Month	Number of bicycles sold
January	30
February	55
March	80
April	150
May	275
June	320

Shop B	
Month	Number of umbrellas sold
January	550
February	480
March	375
April	150
May	225
June	75

a Draw two dot plots to represent each table of data.

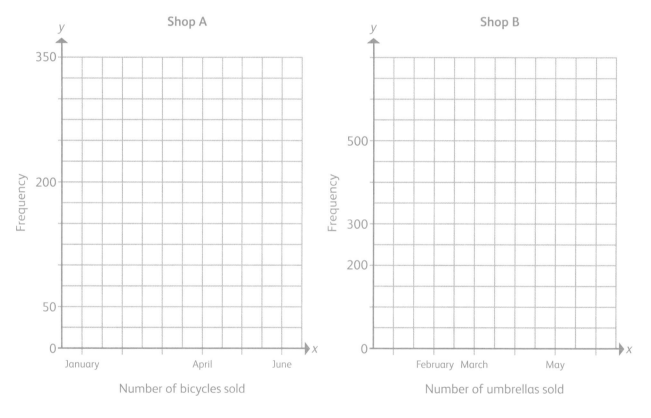

b Describe and explain the patterns you notice in the graphs.
Do you notice any trends?

_____

_____

_____

_____

Frequency charts

1 These are the heights of some trees in a forest:

12 m	31 m	27 m	23 m
32 m	8 m	36 m	26 m
14 m	13 m	27 m	2 m
21 m	29 m	3 m	8 m
28 m	34 m	13 m	36 m
6 m	28 m	34 m	9 m
31 m	3 m	27 m	12 m

Fill in this frequency table.

Height	Frequency
0–5 m	
5 m–10 m	
10 m–15 m	
15 m–20 m	
20 m–25 m	
25 m–30 m	
30 m–35 m	
35 m–40 m	

2 Use the data in question 1 to finish drawing a frequency chart.

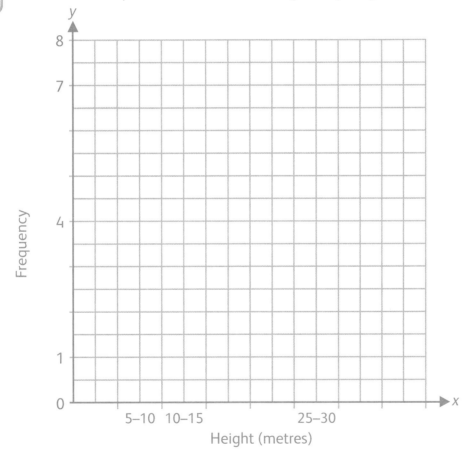

3 What do you notice about the heights of the trees?
Describe the patterns in the data and suggest possible explanations.

_____

_____

_____

Line graphs

1 Use this line graph to answer the questions below.

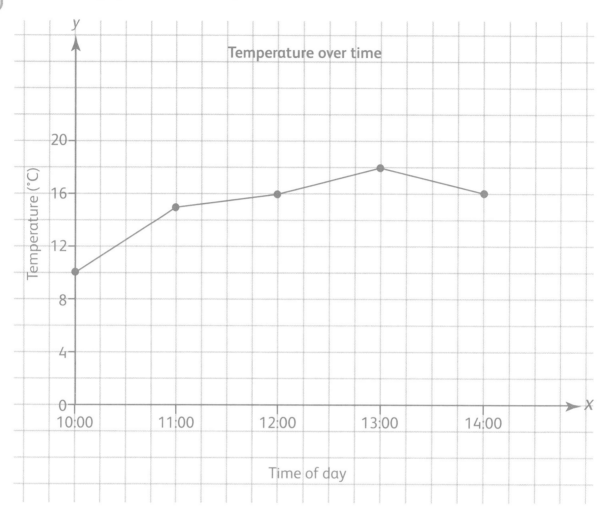

a How many degrees warmer is the temperature at 14:00 than at 10:00?

b Complete this table to show the temperature at each time.

Time	10:15	11:00	12:30	13:00	13:30
Temperature (°C)					

c At approximately what time does the temperature reach 14 °C?

2 Elok fills a sponge with water and then leaves it out to dry.
She records the mass of the sponge every two hours in a table.

Time	0 hours	2 hours	4 hours	6 hours	8 hours
Mass	700 g	625 g	550 g	400 g	275 g

Use the data to draw a line graph below.

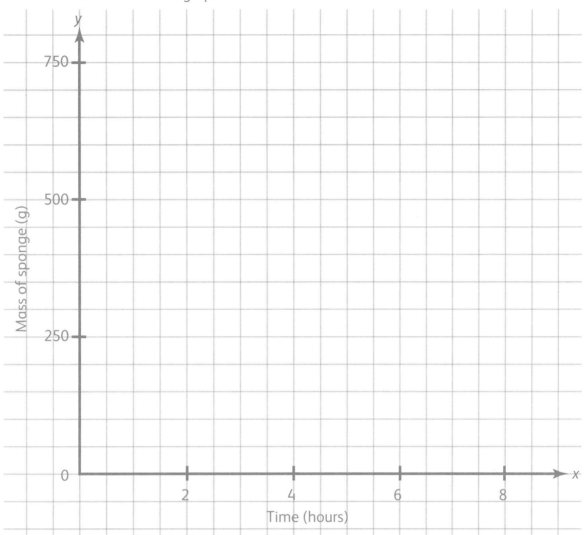

3 Use your line graph from question 2 to answer these questions.

a What is the approximate mass of the wet sponge after 90 minutes?

b For how many hours is the mass of the wet sponge greater than 500 g?

c How many grams lighter is the wet sponge after 6 hours than after 2 hours?

d At approximately what time did the wet sponge weigh 350 g?

Self-check

See how much you know!

 I can do this.

 I can do this, but I need to keep trying.

 I can't do this yet.

What can I do?	😉	🙂	🙁
1 I can read and use bar charts.			
2 I can read and use dot plots.			
3 I can interpret trends and patterns in data shown on dot plots and other graphs.			
4 I can draw and read frequency tables to represent grouped data.			
5 I can draw, read and interpret line graphs.			

I need more help with:

_____

_____

_____

Fractions, decimals, percentages and proportion

Can you remember?

Tick (✓) the shapes that are $\frac{3}{4}$ shaded.

a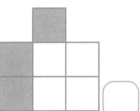

b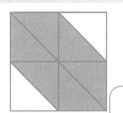

c

d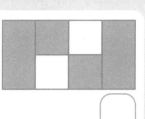

Fractions and division

1 Write the missing numbers or fractions.

a $1 \div 5 = \boxed{} \over \boxed{}$

b $1 \div \boxed{} = \frac{1}{8}$

c $9 \div 10 = \boxed{} \over \boxed{}$

d $1 \div \boxed{} = \frac{1}{6}$

e $19 \div 100 = \boxed{} \over \boxed{}$

2 Guss has three identical bags of sand.
He divides all the sand equally between four containers.

a What fraction of a bag of sand is in each container? $\boxed{} \over \boxed{}$

b What will you draw to **convince** your partner that your answer is correct?

Equivalent fractions

1 Show what you need to multiply both the numerator and denominator by to make these equivalent fractions.

$\frac{1}{4} \rightarrow \boxed{\times 2} \rightarrow \frac{2}{8}$

a

$\frac{3}{5} \rightarrow \boxed{} \rightarrow \frac{6}{10}$

b

$\frac{3}{10} \rightarrow \boxed{} \rightarrow \frac{9}{30}$

2 **a** Draw lines to match the equivalent fractions.

$\boxed{\frac{3}{5}}$ $\boxed{\frac{10}{15}}$ $\boxed{\frac{5}{6}}$ $\boxed{\frac{9}{15}}$ $\boxed{\frac{2}{3}}$ $\boxed{\frac{6}{18}}$

$\boxed{\frac{20}{24}}$ $\boxed{\frac{3}{4}}$ $\boxed{\frac{5}{8}}$ $\boxed{}$ $\boxed{\frac{15}{24}}$ $\boxed{\frac{18}{24}}$

b There is one fraction left over.
Write an equivalent fraction in the blank shape and draw a line to join them.

3 The children are talking about the equivalent fractions $\frac{2}{5}$ and $\frac{8}{20}$.

I noticed that the denominator 20 is a multiple of the denominator 5.

I **conjecture** that all fractions with denominators that are multiples of 5 must be equivalent to a fraction with denominator 5!

Do you agree with Banko?
Explain or show your thinking:

Improper fractions and mixed numbers

 1 Look at the example first. Then convert each mixed number to an improper fraction.

$$3\frac{3}{4} = \frac{\boxed{4}}{4} + \frac{\boxed{4}}{4} + \frac{\boxed{4}}{4} + \frac{\boxed{3}}{4} = \frac{\boxed{15}}{4}$$

a $2\frac{2}{3} = \dfrac{\boxed{}}{3} + \dfrac{\boxed{}}{3} + \dfrac{\boxed{}}{3} = \dfrac{\boxed{}}{3}$

b $3\frac{3}{8} = \dfrac{\boxed{}}{8} + \dfrac{\boxed{}}{8} + \dfrac{\boxed{}}{8} + \dfrac{\boxed{}}{8} = \dfrac{\boxed{}}{8}$

c $2\frac{4}{7} = \dfrac{\boxed{}}{7} + \dfrac{\boxed{}}{7} + \dfrac{\boxed{}}{7} = \dfrac{\boxed{}}{7}$

d $2\frac{1}{2} = \dfrac{\boxed{}}{2} + \dfrac{\boxed{}}{2} + \dfrac{\boxed{}}{2} = \dfrac{\boxed{}}{2}$

2 **a** Complete the table of improper fraction and mixed number equivalents.

Improper fraction	$\frac{3}{2}$	$\frac{7}{4}$	$\frac{7}{5}$	$\frac{11}{8}$	$\frac{11}{4}$
Mixed number					

b Place the mixed numbers on this number line.

0   1   2   3

3 For each pair, tick (✓) the length that is longer.

a $3\frac{5}{8}$ metre ⬚ $\qquad$ $\frac{30}{8}$ metre ⬚

b $2\frac{7}{10}$ metre ⬚ $\qquad$ $\frac{28}{10}$ metre ⬚

c $5\frac{3}{4}$ metre ⬚ $\qquad$ $\frac{22}{4}$ metre ⬚

Fractions as operators

  Find these fractions of amounts. Work across the rows.

a $\frac{1}{2}$ of 75 = [ ] $\frac{1}{2}$ of [ ] = 75 $\frac{1}{2}$ of 15 = [ ]

b $\frac{3}{4}$ of 160 = [ ] $\frac{3}{4}$ of [ ] = 150 $\frac{3}{4}$ of 240 = [ ]

c $\frac{2}{5}$ of 60 = [ ] $\frac{2}{5}$ of [ ] = 60 $\frac{2}{5}$ of 90 = [ ]

  In a row of 200 plants, $\frac{2}{5}$ have red flowers and $\frac{3}{8}$ have yellow flowers.
The remaining plants have no flowers at all. How many of each plant are there?

a [ ] plants with red flowers

b [ ] plants with yellow flowers

c [ ] plants with no flowers

  Pia and Jin have been cutting lengths of ribbon for their kites.

My piece is 45 cm long. I used $\frac{1}{10}$ of my reel of ribbon.

My piece is 40 cm long. I used $\frac{1}{8}$ of my reel of ribbon.

a Who used a larger fraction of their reel of ribbon? _____
b How long was each child's reel of ribbon before it was cut?

Pia: _____ cm   Jin: _____ cm

Adding and subtracting fractions

  Complete these calculations.

a $\frac{6}{10} + \frac{3}{10} =$ [ ]   b $\frac{3}{5} + \frac{3}{10} =$ [ ]

c $\frac{3}{5} + \frac{7}{10} =$ [ ]   d $\frac{\boxed{}}{5} + \frac{7}{10} = \frac{15}{10}$

e $\frac{2}{3} + \frac{5}{6} =$ [ ]   f $\frac{2}{3} + \frac{\boxed{}}{9} = \frac{11}{9}$

Some pizza slices are left over after Sanchia's party.
The adults decide to eat some.
What fraction of each pizza is left over now?
Write the subtraction calculations to match.

	Fraction of whole pizza at the end of the party	Fraction of whole pizza eaten by adults	Fraction of whole pizza left over
a		$\frac{3}{9}$	□/□ − □/□ = □/□
b		$\frac{1}{3}$	□/□ − □/□ = □/□
c		$\frac{3}{4}$	□/□ − □/□ = □/□
d		$\frac{3}{5}$	□/□ − □/□ = □/□

Unit 6 — Fractions, decimals, percentages and proportion

Self-check

See how much you know!

 I can do this.

 I can do this, but I need to keep trying.

 I can't do this yet.

What can I do?	😄	😐	🙁
1 I can solve simple equal share problems with unit fractions, $\frac{3}{4}$ solutions and tenths solutions, for example: 17 bags of rice shared equally across 100 cups.			
2 I can identify, find and represent proper fractions that are equivalent.			
3 I can explain the difference between a proper and an improper fraction.			
4 I can express an improper fraction as a mixed number and vice versa, for example: $\frac{7}{2} = 3\frac{1}{2}$			
5 I can explain why, for example, six apples each cut into four equal parts, results in the same amount: $1\frac{1}{2}$ apples.			
6 I can solve problems involving proper fractions acting as operators, for example: What is $\frac{1}{10}$ of a metre of ribbon?			
7 Given the fraction of a quantity, I can find the whole, for example: $\frac{3}{10}$ of a length of ribbon is 20 cm. How long was the whole ribbon?			
8 I can estimate and add or subtract fractions with the same denominator(s) that are multiples of each other.			

I need more help with:

_____

_____

_____

Can you remember?

Fill in the missing numbers on these calculation paths.

a [23] [× 10] [   ] [× 10] [   ] [÷ 100] [   ]

b [120] [÷ 10] [   ] [× 1 000] [   ] [÷ 100] [   ]

c [268] [× 100] [   ] [÷ 10] [   ] [÷ 1 000] [   ]

Decimal numbers

1 Draw lines to match the numbers.

3.08	zero point three eight
23.6	three point eight
0.38	twenty-three point six
32.6	three point zero eight
3.8	thirty-two point six

2 Write the value of the underlined digit in each number. Work down each column.

a <u>4</u>3.43   [40]   b 50.7<u>6</u> [   ]   c 0.2<u>2</u> [   ]

   43.4<u>3</u> [   ]   5<u>7</u>.76 [   ]   2.<u>2</u>2 [   ]

   143.4<u>3</u> [   ]   557.<u>0</u>6 [   ]   <u>2</u>2.20 [   ]

Place value

1 Describe two different ways to regroup each number. Look at the example.

256	25.6	265.3
• Two hundred and fifty-six ones • 256 ones		

2 Pia and Banko are decomposing numbers. **Critique** their thinking.

Two point three five is equal to two ones, five tenths and three hundredths.

Thirty point zero five is equal to three tens and five tenths.

Make any **improvements**.

_____

_____

3 In each question below, the containers hold a total of 16.75 litres of water. What are some of the possible amounts of water in each container?

a

(ℓ)   (0.25 ℓ)   (ℓ)   (ℓ)

b

(ℓ)   (3 ℓ)   (ℓ)   (ℓ)

c

(ℓ)   (ℓ)   (0.13 ℓ)   (ℓ)   (ℓ)

Rounding to the nearest whole number

 1 Write the previous and the next whole number each time.
Then round to the nearest whole number.

	4.8	6.3	7.5	19.2	123.4
Previous whole number					
Next whole number					
Rounded to the nearest whole number					

2 Guss has rounded each length to the nearest whole metre.

a Which are correct? Tick (✓) them.

b Use the wide boxes to correct any mistakes that Guss has made.

c

Make up two more examples of lengths like these.

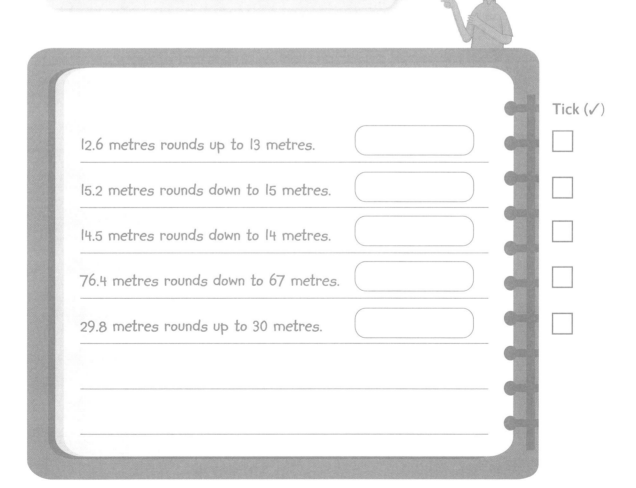

Tick (✓)

12.6 metres rounds up to 13 metres. ☐

15.2 metres rounds down to 15 metres. ☐

14.5 metres rounds down to 14 metres. ☐

76.4 metres rounds down to 67 metres. ☐

29.8 metres rounds up to 30 metres. ☐

Multiplying and dividing decimals by 10 and 100

1 Follow the arrows to take **15.2** from the **Start** through to the **End** by doing the multiplications and divisions.

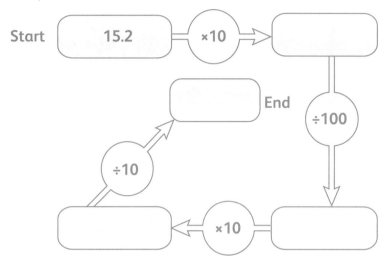

2 The length of a model car is scaled to be 10 times as small as a full-size car.
The length of a model boat is scaled to be 100 times as small as a full-size boat.
Fill in these tables.

	a	b	c
Full-size car (metres)	4.8		5.1
Model car (metres)		0.56	

	d	e	f
Full-size boat (metres)	54		109
Model boat (metres)		0.37	

Patterns and sequences

1 This linear sequence is made with 25c coins.

What is the total value of each term in the sequence? Complete the table.

Term	1	2	3	4	5	8
Total value						

Self-check

See how much you know!

 I can do this.

 I can do this, but I need to keep trying.

 I can't do this yet.

What can I do?			
1 I can say and explain the value of different digits in a numeral, for example: the 4 in 86.54 is 4 hundredths.			
2 I can explain the values of digits in numerals where digits are repeated, for example: the first digit 1 in the number 102.14 represents 1 hundred, whereas the second digit 1 represents 1 tenth.			
3 I can read and write numbers (symbols and words) with up to two decimal places.			
4 I can decompose numbers into the related place value numbers, including decimals, for example: 20.56 = 20 + 0 + 0.5 + 0.06.			
5 I can regroup numbers in a variety of ways, for example, I can express 20.56 as: 20 + 0.56, 2056 hundredths, 2 tens and 56 hundredths.			
6 I can compare and order fractional quantities that are expressed in a variety of forms such as *Order these from smallest to largest*: 1.3, 0.5, $\frac{1}{2}$, 80%.			
7 I can solve problems involving proper fractions acting as operators, for example: What is $\frac{3}{100}$ of 1 kg of flour?			
8 I can find the whole, given the fractional part of a quantity, for example: $\frac{1}{10}$ of a length of ribbon is 20 cm. How long was the whole ribbon?			
9 I can describe ratio or proportion in words, as a fraction or a percentage.			

I need more help with:

_____

_____

_____

Can you remember?

Describe an event for each statement.

a   It is very likely to happen.

_____

_____

_____

b   It is equally likely to happen or not happen.

_____

_____

_____

Equally likely, more likely, less likely

1

| impossible | unlikely | equally likely | likely | certain |

Pia has six boxes with a different object hidden in each one.

Object	A	B	C	D	E	F
Mass	100 g	$\frac{1}{4}$ kg	0.5 kg	150 g	200 g	400 g

She asks Elok to choose one box.
Elok picks a box without looking inside it.
Complete each statement.
The probability of picking a mass that is:

a   lighter than 0.1 kg is _____.

b   a multiple of 100 g is _____.

c   a multiple of 50 g is _____.

d   less than 225 g is _____.

e   _____ is unlikely.

2 Jin uses six 1-digit cards in a box to make these probabilities true.

> The probability of picking an even digit from the box is **likely**.
> The probability of picking a prime number is **equally likely**.
> The probability of picking a square number is **unlikely**.
> The probability of picking zero is **impossible**.

a Write a set of six numbers that Jin could use.

☐ ☐ ☐ ☐ ☐ ☐

b Write a set of six numbers that Jin could not use. Then explain why he cannot use them.

☐ ☐ ☐ ☐ ☐ ☐

_____

_____

Probability experiments

1 To play this game, spin both spinners and add the scores. Do this 10 times or more.

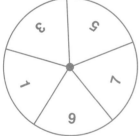
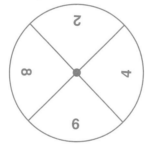

Use a paperclip and a pencil to spin each spinner.

Before you start to play this game, complete these three predictions. I predict that:

a The most common score will be either ☐ or ☐ because

_____ .

b The chance of scoring an even number will be _____ .

c The chance of scoring an odd number will be _____ .

Explain your predictions.

_____

_____

_____

2 Use this table to record your scores as you continue to play the game.

Score	Tally
1–4	
5–8	
9–12	
13–16	
17–20	

a After 10 spins, was your prediction about the most common score correct? _____

b After 20 spins, which is the most common group of scores? _____

c Use the grid below to draw a frequency chart of your results.

3 Invent your own rules. Make some predictions and play with a partner.

You could multiply the two scores or find the difference.
Or, you could double one and add it to the other.

Self-check

 I can do this.

 I can do this, but I need to keep trying.

 I can't do this yet.

See how much you know!

What can I do?			
1 I can recognise when outcomes are equally likely, more likely or less likely.			
2 I can explain why certain outcomes are more or less likely than others.			
3 I can perform probability experiments, make predictions and interpret the results.			

I need more help with:

_____

_____

_____

Unit 9   Calculation

Can you remember?

The number in the centre of each wheel is the answer to the division calculations around it.
Write different division calculations in the empty boxes.

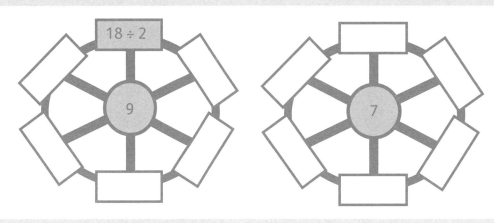

18 ÷ 2

9

7

Addition and subtraction

1   Complete these.

     a   −45 + 90 =

     b   50 − 100 =

     c   −100 + 125 =

     d   −125 + 200 =

     e   15 °C higher than −4 °C is [   ] °C

     f   18 °C lower than 7 °C is [   ] °C

2   Here are the masses of different objects.

> 455 g   676 g   877 g   764 g   1 245 g   2 399 g   1 723 g   2 080 g

Is this statement **always**, **sometimes** or **never** true?
The total mass of four objects will be heavier than the total mass of three objects.

Use estimates to help you. Give example calculations to **convince** others of your answer.

_____

_____

_____

3 Pia subtracts a 4-digit number from another 4-digit number.
She makes an estimate first. Her estimate is 4 500.
a Find the subtraction that Pia is doing and calculate the answer.

9 185 – 4 212   6 127 – 1 997   7 969 – 3 612   5 446 – 2 979

Answer:

Remember to check your answer with my estimate!

b Now choose another calculation from part **a** and make an estimate first.
Complete the calculation and check the answer with a calculator.

Answer:

Adding and subtracting decimal numbers

1 Use rounding to make an estimate first.
Will the actual answers be more or less than your estimates?
Then complete the calculations. Look at the example.

		Estimate	Will answer be more or less?		Actual answer
	4.6 + 6.1	5 + 6 = 11	More	(Less)	10.9
a	9.7 – 2.2		More	Less	
b	10.4 + 5.7		More	Less	
c	13.4 – 4.7		More	Less	
d	18.2 + 3.4 + 5.9		More	Less	

2 Complete these sentences.

a [m] is 2.5 m longer than 5.1 m.

b 9.7 cm is [cm] longer than 6.3 cm.

c The difference between 4.6 ℓ and 8.1 ℓ is [ℓ]

d 0.9 kg heavier than 6.8 kg is [kg]

e 3.2 kg + 4.3 kg + [kg] = 10 kg

3 Arrange the numbers in the diagram so that the total of the shaded square is 20 and the total of each diagonal of four squares is also 20.

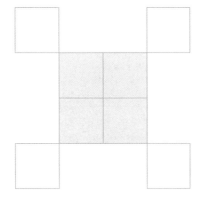

6.3   5.4   4.1   3.7
4.6   5.2   4.8   5.9

Multiplying by a 2-digit number

1 Use an estimate to help you check each calculation. Then solve them.
What methods will you use?

		Estimate	Answer ✓ or ✗	Your calculation
a	323 × 32 = 10 336			
b	54 × 98 = 2 392			
c	254 × 48 = 14 192			

2

a   Complete the table to show the total lengths of the ribbons.

	19 ribbons	21 ribbons	38 ribbons	42 ribbons
Stars				
Stripes				
Spots				

145 cm

218 cm

293 cm

b   How much longer is nine lengths of the ribbon with stripes than nine lengths of the ribbon with stars?   [ ] cm

3

Elok **conjectures** that the product of a 3-digit number multiplied by a 2-digit number will always be larger than the product of a 2-digit number multiplied by a 3-digit number. Do you agree? Use examples to **convince** others of your thinking.
Are there any **generalisations** you can make?

Division

1 Fill in the value of each shaded part when 648 is divided equally.

648

2 Will you round the answers or convert the remainders to fractions?

a 35 tins are arranged in boxes of eight.
How many full boxes are there?

⬚ boxes

b 48 children sit in rows of nine.
How many rows are needed for all the children?

⬚ rows

c The woodcutter cuts a 62 cm length of wood equally into
five short pieces. How long is each piece?

⬚ cm

3 Here are five single-digit numbers:

3	4	5	7	8

a Choose three digits and arrange them in the division calculation so that your
answer has a remainder. Then convert the remainder to a fraction.

 r _____

Answer:

b Choose three digits to find the calculation that gives you the largest remainder.
Choose three more digits for the calculation with the smallest remainder.
Record your calculations and answers.
Convert the remainders to fractions.

Order of operations

1 Complete these calculations.
Remember to think about the order of operations.

a $4 \times 6 \div 2 =$ ⬚

b $100 - 300 + 500 =$ ⬚

c $45 + 6 \times 7 =$ ⬚

d $72 \div 8 - 10 =$ ⬚

e $100 - 81 \div 9 =$ ⬚

f $200 - 25 \times 4 =$ ⬚

2 Banko has 20 marbles.
Pia has 100 marbles. She shares them equally with Banko.
a How many marbles does Banko then have in total?

⬚ marbles

b Complete these number sentences to represent the problem.

⬚ ÷ ⬚ + ⬚

⬚ + ⬚ ÷ ⬚

3 a **Critique** these calculations. Add a tick (✓) to the correct ones.
  i $12 - 4 \times 5 = 40$ _____
  ii $7 \times 5 - 3 = 14$ _____
  iii $40 - 35 \div 7 = 35$ _____
  iv $99 + 8 \times 10 = 1\,070$ _____
  v $500 - 100 \times 3 = 200$ _____

b **Improve** any calculations that are incorrect.

Unit 9 Calculation

Self-check

 I can do this.

 I can do this, but I need to keep trying.

 I can't do this yet.

See how much you know!

What can I do?			
1 I can add a positive integer to a negative integer, such as: $-7 + 180$.			
2 I can subtract a positive integer from an integer where the answer is negative, for example: $45 - 60$.			
3 I can decide whether to work mentally, with jottings or using a formal method, such as: $243 + 171 + 357 = 243 + 357 + 171 = 600 + 171$.			
4 I can use rounding to estimate answers and say when they are correct.			
5 I can compose, decompose and regroup numbers, including decimals, to make calculations easier and more efficient.			
6 I can add or subtract numbers with one decimal place.			
7 I can round one-decimal place numbers to the nearest whole number.			
8 I can use the laws of arithmetic to help me simplify calculations.			
9 I can multiply and divide whole numbers by 10, 100 and 1 000 and explain the answers using place value.			
10 I can estimate and multiply whole numbers up to 1 000 by 2-digit and 1-digit whole numbers.			
11 I can convert remainders into fractions of the divisor when dividing 2-digit numbers by 1-digit numbers.			
12 I can apply skills of rounding and calculating to problems.			
13 I can use the order of operations correctly to carry out calculations where there are no brackets, such as: $3 + 5 \times 2$, $8 + 8 \div 2$, $15 - 10 \div 5$.			

I need more help with:

_____

_____

Can you remember?

Write the coordinates of each vertex in the boxes on the right.

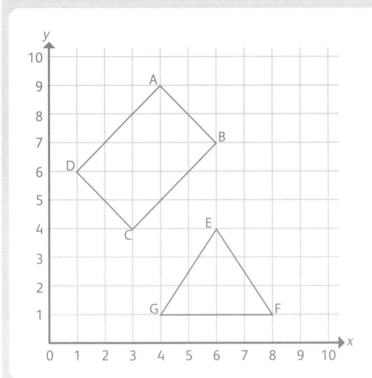

Vertices

A =

B =

C =

D =

E =

F =

G =

Translations

1 Draw the following translations.
Translate shape A:
6 squares right

Translate shape B:
4 squares down

Translate shape C:
3 squares up and
9 squares left.

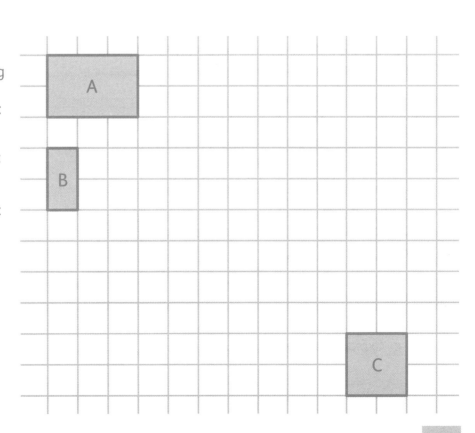

2 Start with shape A.
   a Translate it three squares up and four squares right. Label it as B.
   b Translate shape B five squares right and two squares down. Label it as C.

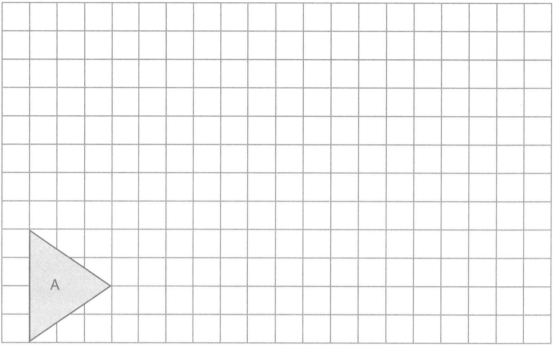

   c Describe the translation needed to move shape C to shape A.

_____

_____

3 Shape X has been translated to a new position.
   a Draw lines to join the translated vertices. One has been drawn for you.
   b Measure the length of each translated line.

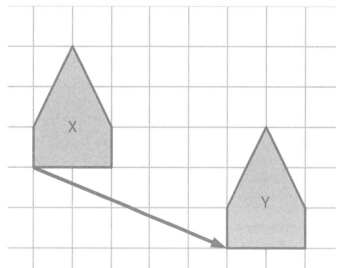

   c Explain what you notice.

_____

_____

a Draw three different translations of shape Z.

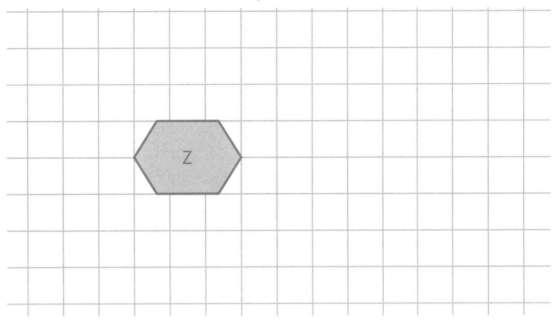

b Explore the lines when you join the vertices of shape Z to the translated vertices. Are the lines always the same length? Are they all parallel? What do you notice?

Shapes on a coordinate grid

a Plot these points on the coordinate grid: (5, 5), (2, 2), (8, 8), (10, 10).

b Follow the same pattern to plot three more points. Write the coordinates here:

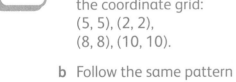

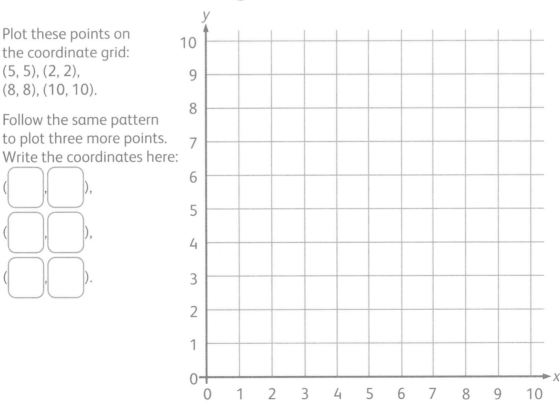

2 A game for two players, A and B.
Take turns to choose two numbers from the list. Then cross them out.

0, 0, 1, 1, 2, 2, 3, 3, 4, 4, 5, 5, 6, 6, 7, 7, 8, 8, 9, 9, 10, 10

Use the two numbers as your coordinates to plot a dot on the grid.
Keep going until you have crossed out all the numbers.
Player A scores a point for every dot inside the rectangle.
Player B scores a point for every dot outside the rectangle.

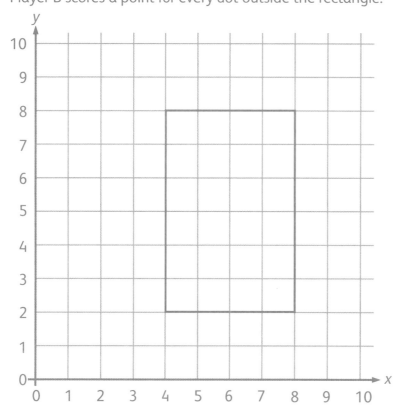

3 Write the missing coordinates for the vertices of the shapes in the grid.

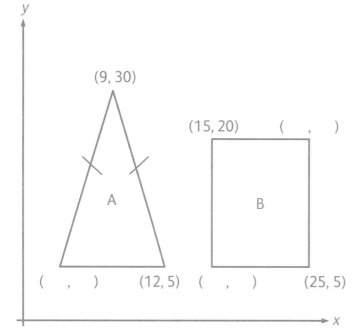

Self-check

 I can do this.

 I can do this, but I need to keep trying.

 I can't do this yet.

See how much you know!

What can I do?			
1 I can translate shapes on a square grid.			
2 I can describe translations on a square grid.			
3 I can use coordinates to give the vertices of 2D shapes.			
4 I can find coordinates on grids without gridlines.			

I need more help with:

_____

_____

_____

Can you remember?

Fill in the table to show mixed number and improper fraction equivalents.

Mixed number	$2\frac{3}{4}$		$1\frac{9}{10}$		$2\frac{5}{6}$	
Improper fraction		$\frac{13}{5}$		$\frac{14}{3}$		$\frac{27}{8}$

Percentages

1 What percentage of each shape is shaded?

a    b    c   d

⬭ %   ⬭ %   ⬭ %   ⬭ %

2 What percentage of each shape is shaded?
Write each percentage as a fraction with denominator 100.

a 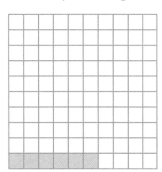   b 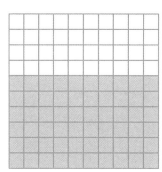   c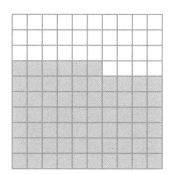

⬭ % = ▭/▭   ⬭ % = ▭/▭   ⬭ % = ▭/▭

3 Banko gets $\frac{75}{100}$ on a test. Sanchia gets 80 % correct on her test.

How many marks more would Banko need, to have the same percentage

correct as Sanchia? ⬭ marks

Equivalent fractions, decimals and percentages

Use the fraction wall to add the missing information.

$\frac{1}{10}$	$\frac{1}{10}$	$\frac{1}{10}$	$\frac{1}{10}$	$\frac{1}{10}$	$\frac{1}{10}$	$\frac{1}{10}$	$\frac{1}{10}$	$\frac{1}{10}$	$\frac{1}{10}$
0.1	0.1	0.1	0.1	0.1	0.1	0.1	0.1	0.1	0.1
10 %	10 %	10 %	10 %	10 %	10 %	10 %	10 %	10 %	10 %

a $0.2 = \dfrac{\boxed{}}{10} = \boxed{}$ %

b $\boxed{}.\boxed{} = \dfrac{\boxed{}}{10} = 50$ %

c $\boxed{}.\boxed{} = \dfrac{6}{10} = \boxed{}$ %

d $0.1 = \dfrac{\boxed{}}{10} = \boxed{}$ %

e $1.0 = \dfrac{\boxed{}}{10} = \boxed{}$ %

f $\boxed{}.\boxed{} = \dfrac{\boxed{}}{10} = 70$ %

2

a   Colour in the flags, as described.

Flag 1	Flag 2	Flag 3
40 % red $\frac{3}{10}$ blue	0.7 red 20 % blue	$\frac{6}{10}$ red 0.2 blue

b   What percentage of each flag is left white?
Write your answers as fractions with denominator 100.

Flag 1: $\dfrac{\boxed{}}{\boxed{}}$ Flag 2: $\dfrac{\boxed{}}{\boxed{}}$ Flag 3: $\dfrac{\boxed{}}{\boxed{}}$

Comparing and ordering quantities

 1 Write <, > or =.

a $\frac{3}{10}$ ◯ $\frac{30}{100}$

b 70% ◯ $\frac{6}{10}$

c 0.8 ◯ 75%

d $\frac{20}{100}$ ◯ 0.2

e 0.9 ◯ 1.2

f 75% ◯ $\frac{3}{4}$

 2 Write these quantities in order from smallest to largest.

| 0.9 | $\frac{40}{100}$ | 50% | 0.2 | $\frac{7}{10}$ |

Smallest _____ Largest

3 Banko says that 70% is greater than $\frac{7}{10}$ because 70 is a bigger number than 7. Explain the mistake Banko has made.

_____

_____

_____

 4 Play this game with a partner. You can use the spinner on this page: straighten one side of a paperclip. Place a pencil in the loop and put the tip of the pencil in the centre of the spinner. Hold the pencil still but spin the paperclip.

Take turns to spin the spinner.

Decide where you will write your number each time.
The aim of the game is to make your statement true.
Think carefully! Miss a turn if you make an incorrect statement.

Player 1: ◯ ◯ % < 0.◯ < $\frac{◯}{10}$

Player 2: ◯ ◯ % < 0.◯ < $\frac{◯}{10}$

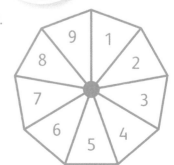

Fractions as operators

1 Find the following fractions of each whole.
Write the matching number sentences.

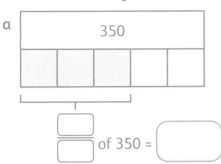

a
350

$\frac{}{}$ of 350 =

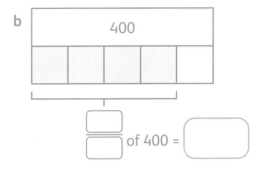

b
400

$\frac{}{}$ of 400 =

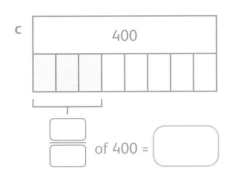

c
400

$\frac{}{}$ of 400 =

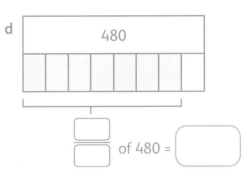

d
480

$\frac{}{}$ of 480 =

2 An architect draws a plan so that all measurements are $\frac{1}{100}$ of real-life sizes.
Complete the table.

	A	B	C	D	E
Real-life measurement	4 metres		15 metres		5 325 centimetres
Measurement on plan		0.12 metres		32.5 centimetres	

3 Elok spends $\frac{1}{4}$ of her pocket money on a comic.
The comic costs $3.

a How much pocket money did Elok have to start with? $

b What fraction of her pocket money does Elok have left? $\frac{}{}$

Ratio and proportion

1 What proportion of the whole group is each shape?

	Triangles	Hexagons	Circles	More than 4 sides	Squares	Not squares
Proportion						

2 Use **for every** to describe the ratios.

| 3 | 8 | 11 | 36 | 15 | 20 | 7 | 49 | 14 | 9 |

Complete the two sentences each time.

a For every _____ odd numbers, there are _____ even numbers.

For every _____ even numbers, there are _____ odd numbers.

b For every _____ 1-digit numbers, there are _____ 2-digit numbers.

For every _____ 2-digit numbers, there are _____ 1-digit numbers.

c For every _____ square numbers, _____ numbers are not square.

For every _____ numbers that are not square, _____ are square numbers.

3 Now make up your own puzzle about ratio by writing some numbers in the box so that:
a For every two multiples of five, there are three multiples of six.
b For every three odd numbers, there are seven even numbers.

4 What proportion of the numbers in question 2 are multiples of three?

Write your answer as a percentage. [ ] %

Self-check

 I can do this.

 I can do this, but I need to keep trying.

 I can't do this yet.

See how much you know!

What can I do?			
1 I can identify simple percentages of shapes and a fraction with denominator 100.			
2 I can give simple equivalences between proper fractions, decimals (one decimal place) and percentages, for example: $\frac{50}{100}$ is $50\% = \frac{1}{2} = 0.5$ $\frac{40}{100} = \frac{4}{10} = 40\% = 0.4$			
3 I can compare and order fractional quantities that are expressed in a variety of forms such as *Order these from smallest to largest*: $1.3, 0.5, \frac{1}{2}, 80\%$.			
4 I can solve problems involving proper fractions acting as operators, for example: What is $\frac{3}{100}$ of 1 kg of flour?			
5 I can find the whole, given the fractional part of a quantity, such as: $\frac{1}{10}$ of a length of ribbon is 20 cm. How long was the whole ribbon?			
6 I can describe a ratio or proportion in different ways; in words or as a fraction or a percentage.			
7 I can describe a situation as a ratio or proportion. For example: In a bag there are four red and six green counters. This becomes: *For every four red counters, there are six green counters*; the proportion of red is $\frac{4}{10}$.			

I need more help with:

_____

_____

_____

Can you remember?

Classify the triangles. Write the letter of each triangle in the correct position on the diagram.

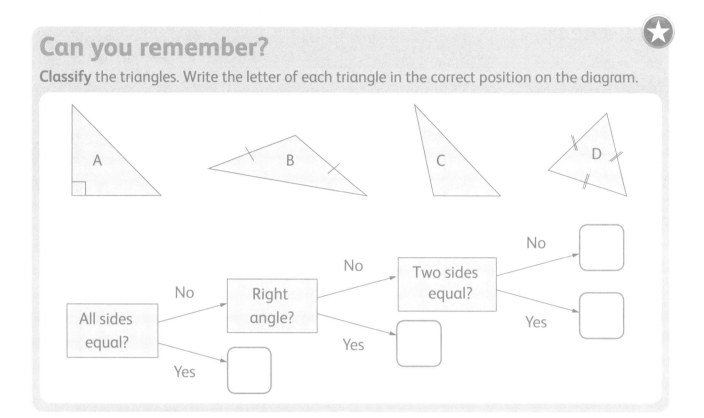

Perimeter and area

1   Draw three different rectangles, each with an area of 20 cm².
Record the length, width and perimeter of each rectangle in the table.

Rectangle	Length	Width	Perimeter
A			
B			
C			

2 Draw three different rectangles each with a perimeter of 200 mm.
Record the area of each shape.

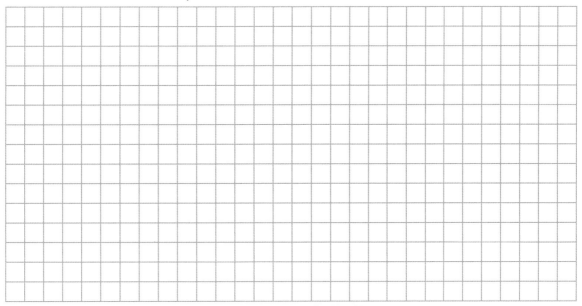

Rectangle	Length	Width	Area
A			
B			
C			

3 Draw a compound shape with the same perimeter as this triangle.

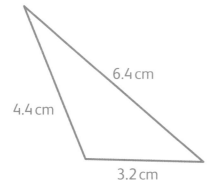

6.4 cm

4.4 cm

3.2 cm

4 Draw a rectangle with the same area as this compound shape.

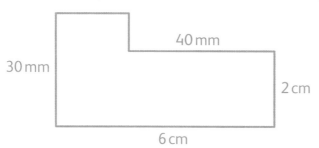

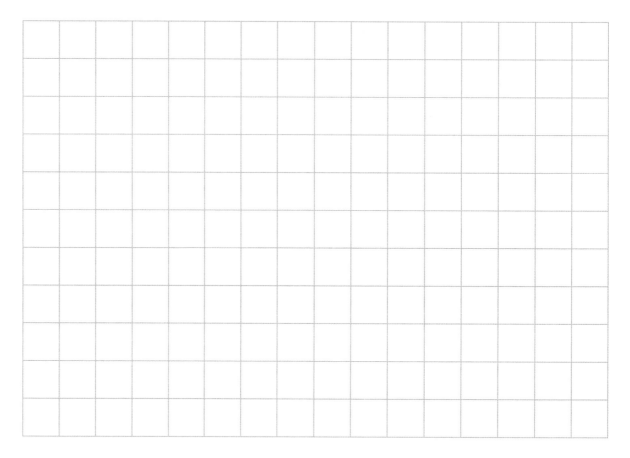

3D shapes

1 Each drawing is a net for an open cube.
Add one face to each, so that it forms a net for a closed cube.

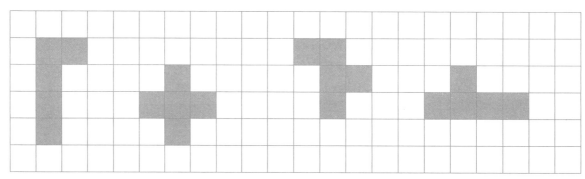

2 Copy these sketches of a cube and other cuboids.

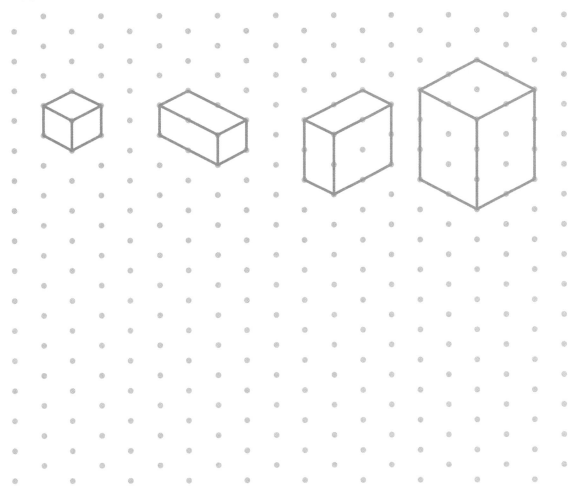

3 Copy these sketches of 3D shapes.
Write the name of each shape under it.

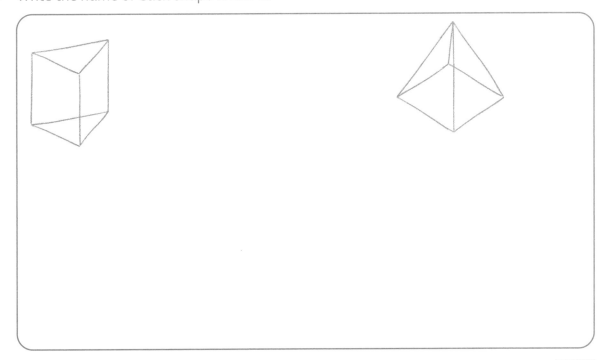

Unit 12   Angles and shapes

Self-check

 I can do this.

 I can do this, but I need to keep trying.

 I can't do this yet.

See how much you know!

What can I do?			
1 I can estimate and measure the perimeter of 2D shapes.			
2 I can reason about shapes with the same area or perimeter.			
3 I can identify nets for open and closed cubes.			
4 I can sketch 3D shapes.			
5 I can identify 3D shapes from different perspectives.			

I need more help with:

_____

_____

_____

Can you remember?

Circle the numbers that round to 10 to the nearest whole number.

| 10.5 | 8.5 | 10.1 | 9.9 | 9.3 | 9.5 | 10.6 |

Square numbers

1 Use two different-coloured pencils. Shade squares to show how the square numbers 4, 9 and 16 are built from consecutive odd numbers.

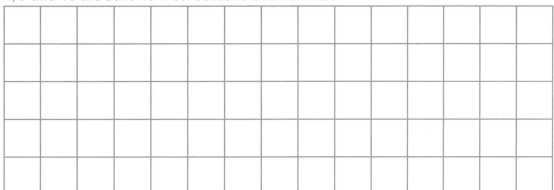

2 Jin is finding out about square numbers and their factors.
He makes this **conjecture**:

I think that all square numbers will have an odd number of factors.

a What do you think? **Critique** Jin's idea.

b Now use what you know about factors to show or explain why 24 is not a square number.

Triangular numbers

1 This is the start of the sequence of triangular numbers:

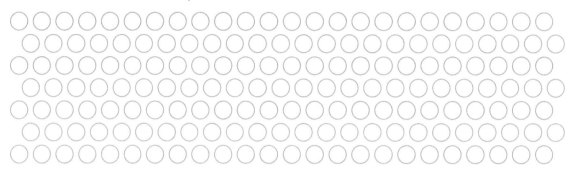

(1), (3), (6), ( ), ( ), ( ), ( )

    a Colour in circles to show the next four triangular numbers in the sequence.
    b Write the numbers in the sequence above.

2 Sanchia is **classifying** some numbers. **Critique** what she has done so far.
Where should she put the remaining three numbers?
Show or explain how you know.

Triangular numbers	Not triangular numbers
28 45	16 25

(21) (36)

(57)

Tests of divisibility

 1 Pia and Jin are talking about some numbers.

> 354 must be divisible by four because the ones digit is four!

> 2356 must be divisible by eight because 56 is divisible by eight!

Both children have made a mistake.
What should they do differently to check their numbers?

_____

_____

_____

_____

 2 a Write each number in the correct section of the Carroll diagram.

 134   2428   1200   3134

428   300   738   3656

	Divisible by 4	Not divisible by 4
Divisible by 8		
Not divisible by 8		

b Write other 4-digit numbers of your own in each section.

c Explain why one section is empty.

_____

_____

Prime numbers

 1 Which of these are prime numbers? Which are composite numbers? Show how you know.

 5   12   11   16

2 Circle the prime numbers in this sequence.

| 7 | 17 | 27 | 37 | 47 | 57 | 67 | 77 | 87 | 97 |

3 Find two or three prime numbers that you can add to make these numbers.
Fill in the table to show the prime numbers you added.

Number	I made it by adding these two or three prime numbers:
5	3 + 2
8	
11	
15	
20	
25	

Self-check

 I can do this.

 I can do this, but I need to keep trying.

 I can't do this yet.

See how much you know!

What can I do?	😉	😐	🙁
1 I can build square numbers in sequence, explaining how to build the next square number from the previous one.			
2 I can identify and say the squares of all digits to 10.			
3 I can use factor pairs to show whether or not a number is square.			
4 I can build the triangular numbers in sequence.			
5 I can show why a number is or is not a triangular number.			
6 I can check and explain why a number up to 100 is prime.			
7 I can check and explain why a number up to 100 is composite.			
8 I can check whether or not a number is divisible by 4 or 8, for example: 3 728 − 28 is divisible by 4 and 3 728 is divisible by 8.			
9 I can explain why the divisibility rules for 4 and 8 work, for example: Take away the last two digits on a number and you are left with a multiple of 100, which is always divisible by 4, so if the last two digits are a multiple of 4, then the whole number is a multiple of 4.			

I need more help with:

_____

_____

_____

Can you remember?

Draw two different shapes with reflective symmetry.

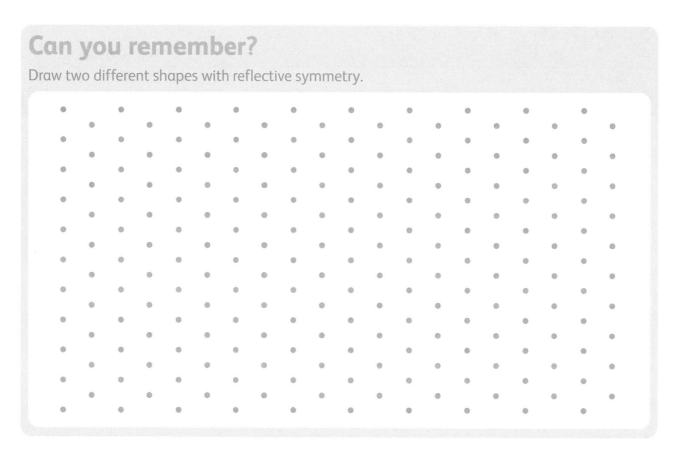

Reflection and translation

1 Draw the reflection of each shape.

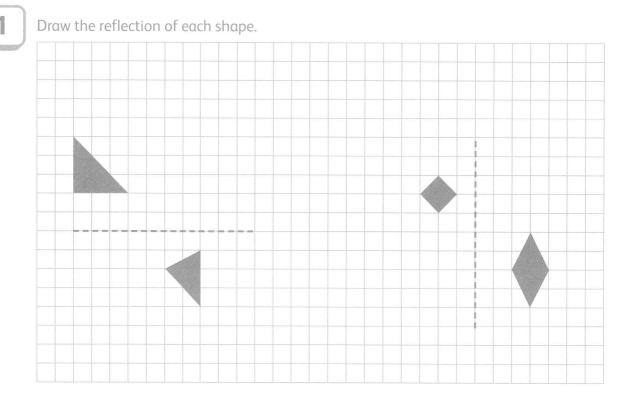

2 Add squares to each pattern, so that it has at least two lines of symmetry.

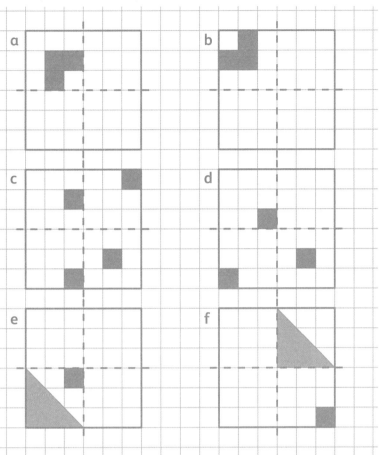

3 Translate each shape as instructed.
Then draw a line of symmetry between the two shapes.

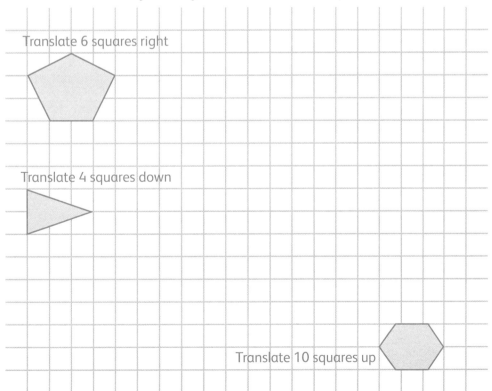

Translate 6 squares right

Translate 4 squares down

Translate 10 squares up

Unit 14 Location and movement

Self-check

 I can do this.

 I can do this, but I need to keep trying.

 I can't do this yet.

See how much you know!

What can I do?			
1 I can translate shapes on square grids.			
2 I can reflect shapes and patterns in two lines of symmetry.			

I need more help with:

_____

_____

_____

Unit 15　Calculation

Can you remember?

Decompose these numbers.

a　2.35 = ⬚ + ⬚ + ⬚

b　3.52 = ⬚ + ⬚ + ⬚

c　3.02 = ⬚ + ⬚ + ⬚

d　5.3 = ⬚ + ⬚

e　0.99 = ⬚ + ⬚ + ⬚

Missing number problems

1　Sanchia and Jin buy a T-shirt and two caps for a total of $25.
Jin gets $9 change when he pays for the T-shirt with a $20 note.

Use the symbols ⬭ and △ to represent the cost of the two items.

Write two number sentences to match the problem.
Use these to find the cost of the T-shirt and one cap.

_____　T-shirt = $ _____

_____　Cap = $ _____

2　Find the mass of the cuboid and a cylinder.

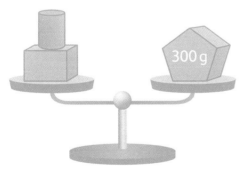
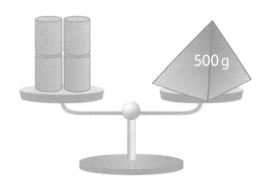

Cuboid = ⬚ g　　　　Cylinder = ⬚ g

Order of operations

 1 Use five numbers once to make each calculation equal 24.

 8   9   12   15   16   18

$30 - \boxed{} \div 3 = 24$ $\boxed{} + 4 \times 2 = 24$ $20 + 32 \div \boxed{} = 24$

$8 \times \boxed{} \div 3 = 24$ $14 - \boxed{} + 25 = 24$

2 Write matching number sentences for the word problems below.
Is there more than one possible number sentence each time?

a Jin has 20 cents more than Guss.
Guss has nine 10 cent coins. How much does Jin have? $\boxed{}$ cents
Number sentence:

_____

_____

b A radio show has 15 songs to play. Each song lasts for four minutes.
It takes 32 minutes to play a certain number of songs.
How many songs do they have left to play? $\boxed{}$ songs
Number sentence:

_____

_____

Multiplication and division

1 Fill in the missing digits to make these calculations correct.

a

1000s	100s	10s	1s
	$\boxed{}$	4	$\boxed{}$
×			7
2	3	9	4

b

1000s	100s	10s	1s
	$\boxed{}$	4	6
×		2	$\boxed{}$
	7	3	8
4	9	2	0
$\boxed{}$	6	5	8

2 The pictogram shows the number of each type of tent at a large campsite.

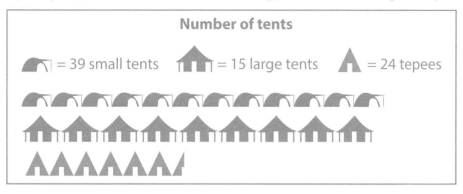

Number of tents

= 39 small tents   = 15 large tents   = 24 tepees

a How many more small tents are there than tepees?

b How many tents in total?

c Tepee tents can sleep up to six people.
   What is the greatest number of people who can sleep in the tepees?

d A school party of 90 children arrives at the campsite.
   A large tent sleeps seven people.
   How many large tents are needed for 90 children?

3 A group of 100 people can fill large tents and tepees with no one left over.
Is this statement true or false?
How will you **convince** others that your answer is correct?

_____

_____

_____

4 Pia and Banko are planting flowers. Each flower costs 99c.

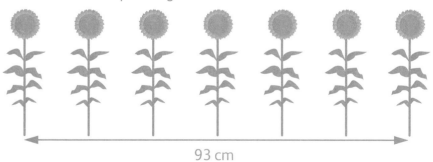

93 cm

The children arrange the flowers so that the distance between each one is the same.

a How much do the flowers cost in total?   cents

b What is the distance between each flower?   cm

5 Guss uses only three digits to make each division calculation.

 6 ☐ ÷ ☐ = 27 r 3

☐☐ 4 ÷ ☐ = 32 r 4

a Which three digits can he use for both calculations?

☐☐☐

b Write the two calculations.

_____

_____

Adding and subtracting decimal numbers

1 Fill in the missing lengths so that each row equals 25.87 cm.

25.87 cm		
a	12.87 cm	
b	18.95 cm	
c	4.76 cm	
d		14.53 cm

2 Answer these questions.

a The total of three numbers is 20.
Two of the numbers are 6.8 and 10.
What is the third number?

b The difference between the lengths of two ribbons is 26.9 cm.
The shorter ribbon is 14.35 cm.
What is the length of the longer ribbon?

3

The table shows the prices of the same board game and computer game at three different shops.

	Board game	Computer game	Total
Kids' Fun	$8.79	$19.79	
Go Game!	$7.95	$20.45	
Let's Play	$9.75	$18.29	

a Write the missing totals in the table.

b How much cheaper is the board game at Go Game! than at Kids' Fun? $

c How much more expensive is the computer game at Go Game! than at Let's Play? $

d Sanchia has $28 and wants to buy both games.
Where should she buy each game and how much money will she have left?

_____

_____

4 Choose a number from box X and one from box Y each time.

X   46.32   27.95   36.83   17.54   **Y**   19.72   36.29   20.68   47.61

a Make three totals that are greater than 60.

b Make three differences that are less than 20.

Multiplying decimal numbers

1 Draw lines to match the multiplications with the same products.

0.7 × 4	1.8 × 3
1.4 × 3	1.4 × 2
0.9 × 6	0.7 × 6
1.8 × 4	1.4 × 6
1.2 × 7	0.9 × 8
	0.9 × 9

2 The mass of a bag of pebbles is 4.6 kg.
The mass of a bag of sand is 12.3 kg.
What is the total mass of four bags of pebbles and three bags of sand? ⬚ kg

3

I **conjecture** that when I multiply 0.6 by each number from 1 to 10, all the products will be ten times as small as the products in the multiplication table of six!

a Do you agree? How will you **convince** others of your decision?

b What **conjectures** such as Jin's can you make of your own?

Unit 15 Calculation

Self-check

 I can do this.

 I can do this, but I need to keep trying.

 I can't do this yet.

See how much you know!

What can I do?	😊	😐	🙁
1 I can recognise and use symbols or shapes to represent unknown quantities.			
2 I can use related facts and inverse operations to help with some missing number problems.			
3 I can use my knowledge of the order of operations correctly to carry out calculations with no brackets, such as: $3 + 5 \times 2$, $8 + 8 \div 2$.			
4 I can use my understanding of place value when multiplying and dividing with larger numbers.			
5 I can estimate and multiply whole numbers up to 1 000 by 1-digit or 2-digit whole numbers.			
6 I can estimate and divide whole numbers up to 1 000 by 1-digit whole numbers.			
7 I can convert remainders into fractions of the divisor when dividing 2-digit numbers by 1-digit numbers.			
8 I can compose, decompose and regroup numbers, including decimals to make calculations easier and more efficient.			
9 I can add or subtract numbers with one or two decimal places.			
10 I can estimate and multiply numbers with one decimal place by 1-digit whole numbers.			
11 I can round numbers with one decimal place to the nearest whole number.			
12 I can apply my skills of rounding and calculating to problems.			

I need more help with:

_____

_____

_____

Can you remember?

Complete these percentages.

a 50 % of 40 = [ ]

b 10 % of 40 = [ ]

c 50 % of 80 = [ ]

d 10 % of 80 = [ ]

e 50 % of 800 = [ ]

f 10 % of [ ] = 800

Mode and median

1 Circle the median in each set of numbers. Write the median and mode for each set.

a 1, 2, 3, 4, 4   Median [ ]   Mode [ ]

b 0, 0, 1, 1, 1   Median [ ]   Mode [ ]

c 1, 2, 3, 4, 5, 5, 6   Median [ ]   Mode [ ]

d 11, 9, 7, 7, 6, 5, 3, 1, 0   Median [ ]   Mode [ ]

2 Choose a book. Pick any page. Record the number of letters in 19 words in the top row.

In the second row, re-order the numbers from smallest to largest.

What is the median and mode for your set of data?   Median [ ]   Mode [ ]

3 Choose a different page in the book. Repeat the investigation.
Before you record the number of letters in 19 words, complete the prediction.

I predict that the mode and median will be _____

because _____

Complete the investigation to check your prediction.

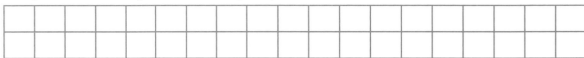

Actual median [ ]   Actual mode [ ]

4 Now choose a page in a different book. Repeat the investigation.
Before you record the number of letters in 19 words, complete the prediction.

I predict that the mode and median will be _____

because _____

Complete the investigation to check your prediction.

Actual median ⬭　　　Actual mode ⬭

Proportion of the whole

1 Use the waffle diagram below to show the data from this frequency table.

Favourite colour	Frequency	Percentage
Red	5	10%
Blue	6	12%
Green	2	4%
Yellow	25	50%
Black	12	24%

2 a Use a pencil and paperclip to spin this spinner 20 times.
Collect the scores and use the waffle diagram to show the results.
Choose a colour for each score.
Write it in the key.

Key:

	1
	2
	3

b Explain the results shown by your waffle diagram.

_____

_____

_____

3 Use your waffle diagram to help you complete this frequency table for your results.

Score	Frequency	Percentage
1		
2		
3		

Unit 16 Statistical methods

Self-check

 I can do this.

 I can do this, but I need to keep trying.

 I can't do this yet.

See how much you know!

What can I do?			
1 I can find the mode of a set of data.			
2 I can find the median of a set of data.			
3 I can use a waffle diagram to show a set of data.			
4 I understand how a waffle diagram shows the proportion of a whole.			

I need more help with:

_____

_____

_____

Unit 17 — Fractions, decimals, percentages and proportion

Can you remember?

Complete these:

a $\frac{3}{5} + \frac{\square}{\square} = \frac{7}{10} + \frac{\square}{\square} = 1$

b $\frac{\square}{\square} + \frac{5}{8} = \frac{\square}{9} + \frac{\square}{9} = 1$

Greater than, less than, equal

1 Circle the larger quantity each time.

a 0.5 $\frac{60}{100}$

b 30 % $\frac{2}{10}$

c 100 %   1.1

d $\frac{9}{100}$ 10 %

2 Order the quantities from smallest to largest by writing the shape names on the line below.

$\frac{70}{100}$ (hexagon)   30 % (circle)   1.3 (square)

100 % (rectangle)   0.5 (triangle) $\frac{9}{10}$ (pentagon)

Smallest _____ Largest

3

I am thinking of a percentage. I can write it as a number of tenths. It is larger than the fraction $\frac{2}{5}$ but smaller than the decimal number 0.8.

What could Sanchia's percentage be? Write three possible answers.

_____

_____

Adding and subtracting fractions

1 Pia draws a diagram to show how to add $\frac{2}{5}$ and $\frac{3}{10}$.

$\frac{1}{5}$		$\frac{1}{5}$					
					$\frac{1}{10}$	$\frac{1}{10}$	$\frac{1}{10}$

$\frac{8}{10}$

Critique Pia's work. Use the space below to draw any **improvements** she should make.

2 Work out these calculations.

a $\frac{3}{4} + \frac{3}{8} = \dfrac{\boxed{}}{\boxed{}}$

b $\frac{3}{4} - \frac{3}{8} = \dfrac{\boxed{}}{\boxed{}}$

c $\frac{2}{3} + \frac{5}{12} = \dfrac{\boxed{}}{\boxed{}}$

d $\frac{11}{12} - \frac{2}{3} = \dfrac{\boxed{}}{\boxed{}}$

3 Guss cycles $\frac{7}{8}$ km. Sanchia cycles $\frac{3}{4}$ km further than Guss.

Elok cycles a distance that is $\frac{1}{2}$ km shorter than Guss.

How far did Sanchia and Elok cycle?　　Sanchia: $\dfrac{\boxed{}}{\boxed{}}$ km　　Elok: $\dfrac{\boxed{}}{\boxed{}}$ km

Multiplying and dividing unit fractions by a whole number

1 What calculations are represented here?

a

$\dfrac{\boxed{}}{6}$

$\dfrac{\boxed{}}{\boxed{}} \times \boxed{} = \dfrac{\boxed{}}{\boxed{}}$

b $\dfrac{1}{\boxed{}}$

$\dfrac{1}{\boxed{}} \div \boxed{} = \dfrac{1}{\boxed{}}$

2 Complete these calculations.

a $\frac{1}{4} \times 2 =$ ☐

b $\frac{1}{4} \div 2 =$ ☐

c $\frac{1}{5} \times 3 =$ ☐

d $\frac{1}{5} \div 3 =$ ☐

e $\frac{1}{6} \times 4 =$ ☐

f $\frac{1}{6} \div 4 =$ ☐

3 Pia takes $\frac{1}{3}$ of a sheet of paper and divides it into two equal pieces.

She takes another $\frac{1}{3}$ and divides it into three equal pieces.

Pia uses the final $\frac{1}{3}$ and divides it into four equal pieces.

What fraction of the whole sheet of paper does each of her smaller pieces represent?
Sketch diagrams and write divisions to **convince** others of your answers.

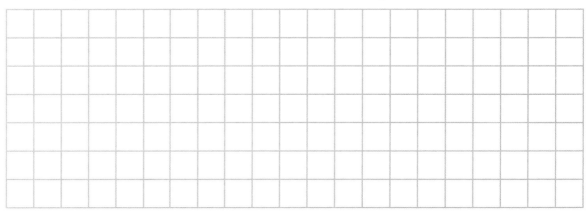

4 The running track at the local park is $\frac{1}{8}$ km long.

Jin runs the length of the track seven times.
What fraction of a kilometre did he run in total?

$\frac{☐}{☐}$ km

Ratio and proportion

1 Colour the beads so that the ratio of red to yellow beads is 2 : 5.

2 Answer the questions about these daisies.

a What is the ratio of darker daisies to lighter daisies in this group of daisies?

b What is the ratio of lighter daisies to darker daisies?

c What proportion of the daisies is darker?

d What proportion of the daisies is lighter?

e Try to give your proportions as percentages.

3 Jin has some red, green and blue marbles in a jar.

The proportion of red marbles is $\frac{3}{8}$.

The proportion of blue marbles is $\frac{1}{4}$.
There are six green marbles.

a How many marbles are in the jar in total?

b How many marbles are red?

c How many marbles are blue?

Unit 17 Fractions, decimals, percentages and proportion

Self-check

 I can do this.

 I can do this, but I need to keep trying.

 I can't do this yet.

See how much you know!

What can I do?			
1 I can give simple equivalences between proper fractions, decimals (one decimal place) and percentages, for example: $\frac{50}{100}$ is $50\% = \frac{1}{2} = 0.5$ $\frac{40}{100} = \frac{4}{10} = 40\% = 0.4$			
2 I can identify simple percentages of shapes and a fraction with denominator 100.			
3 I can compare and order fractional quantities expressed in a variety of forms such as *Order these from smallest to largest*: 1.3, 0.5, $\frac{1}{2}$, 80%.			
4 I can estimate and add or subtract fractions with the same denominator(s) that are multiples of each other.			
5 I can estimate, multiply and divide unit fractions by a whole number, for example: $\frac{1}{3} \times 5$ or $\frac{1}{3} \div 5$			
6 I can describe a situation as a ratio or proportion. For example: In a bag there are four red and six green counters. The ratio of red to green counters is $4:6$. The proportion of red counters is $\frac{4}{10}$.			
7 I can describe a ratio or proportion as percentages (multiples of 10). For example: In a bag there are four red, three green and three blue counters. The ratio of red to green to blue counters is $4:3:3$. The proportion of red counters is $\frac{4}{10}$, which is the same as 40%.			

I need more help with:

_____

_____

_____

Can you remember?

a Circle every Thursday on the calendar.

b The date of the last Friday in March is: _____

c Write the date of the first Friday in May: _____

d Maris swims every Thursday in May.

These dates are: _____

April						
S	M	T	W	T	F	S
					1	2
3	4	5	6	7	8	9
10	11	12	13	14	15	16
17	18	19	20	21	22	23
24	25	26	27	28	29	30

Time zones

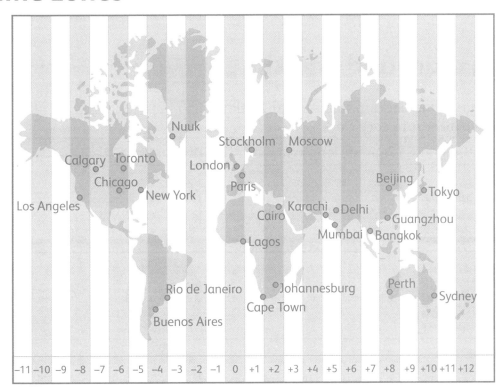

1 Look at the world map above. Find pairs of cities with these time differences:

a _____ is 4 hours ahead of _____

b _____ is 12 hours ahead of _____

c _____ is 19 hours behind _____

d _____ is 9 hours behind _____

2 It is 03:45 in Sydney. Draw hands on the clocks to show the time in each city.

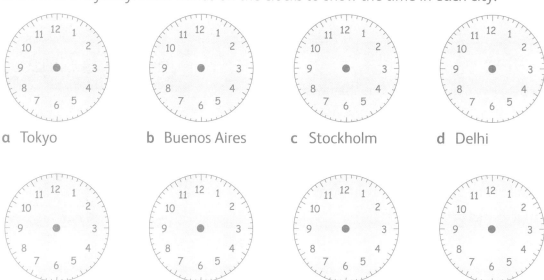

a Tokyo   **b** Buenos Aires   **c** Stockholm   **d** Delhi

e Johannesburg   **f** Guangzhou   **g** Nuuk   **h** Ontario

Calculating start and end times

1 Draw the hands of the clocks to show the arrival time.
Write **a.m.** or **p.m.** next to each clock.

	Start time	Journey length	Arrival time	
a	**10:55**	20 minutes		
b	**08:40**	$\frac{1}{2}$ hour		
c	**23:55**	1 hour and 10 minutes		
d	**11:55**	70 minutes		
e	**03:30**	24 hours and 45 minutes		

2

Play this game with a partner.
The aim of the game is to collect hours, halves or quarters.
You can choose to add on 10 minutes, 20 minutes or 25 minutes with each turn.
- If you make a time that is **quarter past** or **quarter to**, you score **1 point**.
- If you make a time that is **half-past** or **o'clock**, you score **2 points**.
- If you make a time that is **12 o'clock** you score **5 points**.

Take turns to add a time and draw the hands on the next clock.
The winner is whoever scores the most points when you reach the last clock.

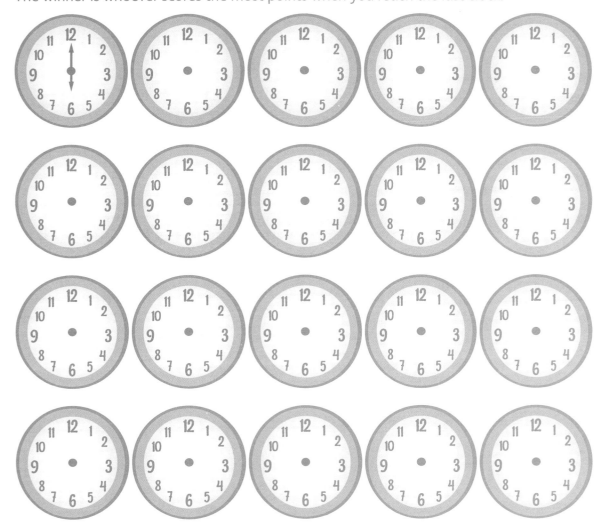

Use this table to write the scores.

Player A score		Player B score	

Unit 18 Time

Self-check

See how much you know!

 I can do this.

 I can do this, but I need to keep trying.

 I can't do this yet.

What can I do?			
1 I can work out times in different time zones.			
2 I can find start and end times for different durations.			
3 I can write time in 12-hour and 24-hour clocks.			
4 I can recognise and use time written as a decimal.			

I need more help with:

_____

_____

_____